Anonymous

A Peep into the House You Live in

An Essay on the Causes of Some of the Ailments to Which Flesh Is Heir and

on the Value of Mechanical Appliances in Their Treatment

Anonymous

A Peep into the House You Live in
*An Essay on the Causes of Some of the Ailments to Which Flesh Is Heir and on the
Value of Mechanical Appliances in Their Treatment*

ISBN/EAN: 9783337853488

Printed in Europe, USA, Canada, Australia, Japan

Cover: Foto ©berggeist007 / pixelio.de

More available books at **www.hansebooks.com**

A PEEP

Into the House You Live In.

AN ESSAY

ON THE

CAUSES OF SOME OF THE AILMENTS TO WHICH FLESH IS HEIR;

AND ON THE VALUE OF

𝔐echanical 𝔄ppliances in their Treatment.

PUBLISHED BY

THE BANNING TRUSS AND BRACE COMPANY,

Office, 704 Broadway, above Fourth Street, N. Y.

PRICE 10 CENTS.

CONTENTS.

For "List of Illustrations" see inside of last page of cover.

THE MECHANICAL NATURE

OF SOME OF THE

CHRONIC MALADIES,

AND

The Advantages of Mechanical Appliances in their Treatment.

"A COMMON SENSE ESSAY."

By THE BANNING TRUSS & BRACE COMPANY OF NEW YORK.

It is apparent to even the most casual observer that there exists a large class of chronic diseases which have to this day steadily resisted every established method for their treatment and cure.

Among these, spinal, digestive, nervous, vocal, pulmonary and female weaknesses stand prominent, and have a representative, to say the least, in a clear majority of families. Startling as this proposition may seem, it is nevertheless a *fact*, staring the practitioner in the face and defying contradiction.

And with a view to elucidating the causes of the ill success in their management, and how it may be that mechanical support when physiologically constructed or applied may, by itself or in connection with proper medical treatment, cure the same by removing the *mechanical cause*, your attention is invited to the following common sense propositions in mechanical pathology—a pathology which has been much *overlooked*, but the importance of which cannot be *overestimated*, when we consider the large class of diseases that are acknowledged to be purely mechanical, to say nothing of that much larger class which, though mechanical, a defective pathology has not heretofore ranked as such.

Proposition I.—That man, materially considered, is a machine—from the grossest fibre to the finest cell—of a primary, definite, and accurate character, and as such is under the control of mechanical law, a change in its actions always resulting from any local or general departure from the primary and accurate position as in other machines. Take, for instance, that complicated and beautiful combination of many parts, the lever watch. If the slightest change takes place in the bearing or position of the smallest spring or wheel, its action upon its neighbor is rendered imperfect, and this imperfection is carried through each succeeding part of the delicate machine, affecting the accuracy of the result as well as rapidly impairing the nice adjustment necessary to make it go at all.

Proposition II.—The internal organs or viscera are as much under the law of definite position as are the bones, and a functional derangement will follow a displacement in one case as well as in the other.

Proposition III.—This definite position and relation of the internal organs consists in a *packed, braced,* and *supported* state from *below,* and not in a *suspended* or pendent state from above. We take the *broad ground* that in the whole normal human body there is not a single *suspended organ.* That this *must* be so is self-evident. These organs are nervous, tender and fragile, easily torn, easily injured, and if allowed to simply swing to and fro from their moorings they would be as liable to damage (to use a homely metaphor) as a satchel of eggs hung up in a stage coach.

The surrounding elastic abdominal walls are the consummative agencies in maintaining this supported and primary state of the vital parts. The abdominal walls are elastic only in proportion as the abdominal and spinal muscles are tensed and energetic—and these can only be so when the body is thoroughly erect; and when these muscles are active and strong, and the median plane of the pelvis is thus rendered oblique instead of horizontal, the visceral weight is thereby thrown forward upon the lower abdominal wall, which, acting under the lowest of the viscera, causes it in its turn to become the aggressor of its superior neighbor, and so on to the top of the pile;

BANNING TRUSS & BRACE CO., 704 Broadway, N. Y.

BANNING TRUSS & BRACE CO., 704 Broadway, N.Y.

by which the lower abdomen becomes comparatively small, the waist and stomach plump, the chest short and broad, all by virtue of the consecutive series of supports from the base to the apex, resting upon the muscles below, which act much like the springs of a carriage, and give us the natural or supported state.   See Fig. 1.

A A, Front line, showing that the big toe, pubis, and tip of the nose, are always in line when the body is erect.   B B, Line, passing through the spine at the neck, and at the hip, knee, ankle joints; and D, showing that these points are in line when the body is erect, and that D is the body's centre of gravity; C C, Posterior line, showing that the back of the head, shoulders, hip, and heel, are also in line, and

that when the body is erect, D is much in advance of this line.   K K and I I, Lines running in the exact direction of the advancing and retreating directions of the Spine, crossing each other and B B at D, and proving mathematically that D is the body's centre of gravity.

This figure shows that the big toe and the nose are in line; even when the hollow of the back has receded from the middle line, beyond line C, and behind the shoulders; that drooping and round shoulders are produced by a retreating motion at D (for, if such were not the fact, the perpendicular lines, in both

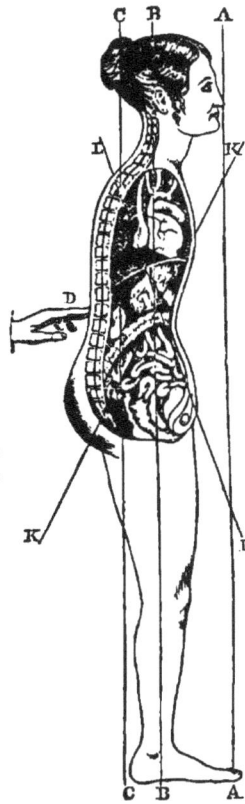

the figures, would not touch the head and feet at the same points), and, that no style of support, designed for straightening the form, can accomplish its object, unless so constructed as to push forward the receded point at D, and bring it again in contact with the line B.   See Fig. 1.

FIG. 1.—Side view of the Erect Posture, with natural upward and inward bearing of the internal organs.

FIG. 2.—Side view of the Drooping Posture, with internal organs suspended and compressed.

Proposition IV. That the effects of a departure from the natural or supported state to the unnatural or pendent state are two-fold: First, the effects of pressure upon the parts below, **and** second, the effects of the change from a supported to a pendent state upon the parts above. For this unnatural state, see Fig. 2.

And first let us consider the

Effects of Pressure on Parts Below.

EFFECT UPON THE NERVES.

The nerves, which conduct the nerve fluid or force generated by the spinal marrow or brain, are thrown out from each of the vertebræ, and clinging close to the bony walls, pass down into the limbs ; they tell us when we tread on an injurious substance ; how far we have raised our feet;—in fact they are what makes us sentient beings. When, as in the unnatural state, there is a pressure placed on them they cannot conduct the fluid freely, the action of the limbs becomes clogged—a sense of weight seems to hang about them—even a straw will cause stumbling, and the ability to exercise becomes greatly diminished. Then follows that dreadful sensation of " gnawing, aching, tearing pain " in the back—for which blisters are applied, and often actual cautery and the moxa ; but the difficulty is not at that point, and the local treatment touches the effect, not the cause of the trouble, which is down in the sacral curve, where the nerves, burdened by the weight of the viscera, are crowded against the bony walls, and are struggling to convey their subtle influences through the restricted passages which obstruct their energies.

EFFECT UPON THE ARTERIES, CAUSE OF COLD AND CLAMMY FEET AND LIMBS.

We see the grand Iliac artery after its distribution in the pelvis, passing off to its remote ramifications ; it too is compressed, and the nourishing blood it conducts, **and** which gives warmth and fulness to the limbs, is obstructed. The feet become cold and clammy. This explains why it is that some people always have cold feet even in July.

EFFECT IN CAUSING VARICOSE VEINS AND DROPSY.

The veins in carrying the blood back to the heart for purification against the force of gravity, meet with this mechanical obstruction in the pelvis, and the blood is thrown back, and varicose veins are the result. For this, elastic stockings and bandages are usually applied; but they do not meet the case, which consists, as before shown, in the pressure upon them in the pelvic cavity. Remove the cause by taking away the weight, and the effect will disappear of itself. The blood is composed of three-fourths water; its red appearance is due mainly to the presence in it of the red corpuscles; while this obstruction exists, the heart, like a stationary engine, is forcing the blood through the arteries, which convey it back through the veins, the result of which is that its watery portion oozes through their minute capillaries, and a dropsical condition ensues.

EFFECT UPON THE RECTUM OR LARGE BOWEL—CONSTIPATION —HEMORRHOIDS OR PILES, AND PROLAPSUS ANI, OR PROTRUSION OF THE BOWEL.

The small bowels, instead of being properly elevated, now press with great force on that part of the backbone which juts forward, and where the large bowel (rectum) dips down into the pelvic or lower cavity, thus imposing a mechanical obstruction to the regular descent of the fæces and producing an accumulation of alvine matter. The natural secretions of the bowel are also absorbed by the hard and heated condition of the mass within, which is constantly irritating the inner membrane and nerves.

In a natural and healthy condition of the body; the contents of the abdomen, whenever we step or fall, descend upon the elastic muscles, which instantly distend, and then in like manner react, throwing themselves and the superincumbent bowels back to their former position. This any one can test by placing his hand upon the lower abdomen, and stepping heavily. He will feel the vibration or reaction of which we speak.

Now, two purposes are subserved by this arrangement. It tends to preserve the organs from injurious contact, by giving them, always, a gentle movement, and it stimulates the bowels to motion and secretion. But how different is the condition of the bowels, when they rest upon the lower abdomen and the bones in that vicinity? Lying below the axis of muscular contraction, and being perfectly inert and totally deprived of their usual facilities for action, they must often press downward with great force. The most obstinate cases of costiveness are thus produced. The whole intestinal tube becomes preternaturally distended, causing bad breath, loss of appetite, and otherwise deranging the stomach. The circulation of the blood is also mechanically obstructed, inducing stupor, sleepiness, ennui, and a host of other symptoms. At length, nature rouses herself to throw off the foreign mass by which she is thus opposed. But she labors under great disadvantages in the performance of this extraordinary task. The bowel is very tender and its contents dry and hard; but by the exercise of unusual strength and force, the work is accomplished. Yet what is the effect? Why, the inner membrane being naturally much larger than the outer, and lying in folds, the dry mass pushes these folds before it, and leaves them at the outlet, where they are held by the contraction of the sphincter muscle, which covers the aperture, until hard and very sore tumors are formed upon the parts. At first, the membrane may be drawn back before the

BANNING TRUSS & BRACE CO., 704 Broadway, N. Y.

sphincter muscle contracts upon it ; but it soon loses its tone, becomes congested and relaxed, and prolapsed at every stool, by the slightest straining. This painful state is greatly promoted by the pressure from above upon the lower bowel, continually bearing it down and increasing the tendency to its inversion. These distressing stools are eventually succeeded by discharges of blood, when we have as genuine a case of piles as can be imagined, which, by the way, is of itself a most distressing malady, though it seldom exists without constipation, as may readily be inferred from the preceding remark. Indeed, no less than eight out of every ten cases of piles, are induced by constipation of the bowels.

Now, is it possible for *medicine* of any kind to effect anything more than temporary relief in such afflictions as these, without first restoring the mechanism of the disordered parts? Surely not. Let us then see to it, that the only rational mode of treating these vexatious and distressing maladies, namely, the elevation and support of the viscera by judicious mechanical means, is at once adopted.

EFFECTS UPON THE URINARY BLADDER. (URINAL INCONTINENCE AND RETENTION—AFFECTION OF PROSTATE GLAND.)

The Urinary Bladder, when compressed by the descent of the bowels, is of course unable to retain its contents a proper length of time, in which case there will be a frequent desire to void the same. This is termed *incontinence*. It is not very dangerous, but exceedingly troublesome and inconvenient. Sometimes an angle is also produced in the *neck* of the bladder, thus mechanically obstructing the passage of the urine, and inducing oftentimes the most terrible and fatal disasters. It is true that there are other causes for this malady, requiring internal treatment. But when the difficulty is purely *mechanical*, which is generally the case, how many pumpkin seeds, how much Harlæm oil, or Spirits of Nitre,

8

will it require to relieve the suppression? Or, in cases of incontinence thus engendered, how much Tinct. Lyttea, Balsam Copaiva, Bitters, &c., will remove this thirty or forty lbs. of visceral weight and thus remedy the matter?

FIG. 3.

FIG. 4.

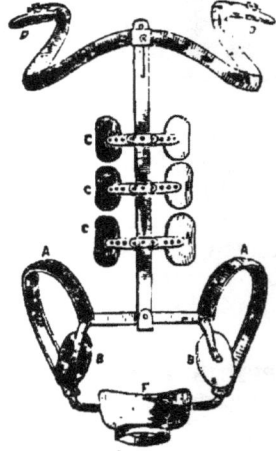

The bladder evidently needs no medicine. Let but the oppressive and superincumbent weight be removed, and all will be well. The patient will then be saved from the annoyance of the catheter and the disgusting use of nostrums. Let no one think of internal dosing without understanding the form and habits of the patient. Let him see that the abdomen is not pendulous at its base, and relaxed in its upper region. Let him be fully assured that the difficulty does not require mechanical aid; for many cases, both of incontinence and retention, after resisting all other treatment, have yielded to the application of the Body-Brace or abdominal and

Fig 3 represents the simple Body-Brace, generally successful; but when not so, the corresponding attachments are required.

Fig. 4 represents Abdominal and Spinal Shoulder-Brace. A A, arches of the mainspring passing above each hip-bone, avoiding all pressure on the bones, nerves, or blood vessels, and furnishing the power of the instrument. B B, pads supporting glutei muscles on either side. C C, aggressive supporting saddles, to either side of the dorso-lumbar spine (in ordinary cases only the lower saddle to be used), holding that portion of the spine properly forward. D D, spring support, resting anteriorly upon the heads of the humeri. F, front pad, elevating abdominal viscera.

The combined action of all is, to elevate the lineal viscera, sustain the lumbar spine in its vertical position, and poise the superior trunk behind the spinal axis.

BANNING TRUSS & BRACE CO., 704 Broadway, N. Y.

spinal shoulder Brace. The most troublesome inconti‑ nence has often been relieved; by merely lifting up the bowels with the hands, so as to change the shape of the abdomen. In retention also, relief has often been afforded by simply turning the sufferer on his face.

Although so dissimilar in their phenomena, both in‑ continence and retention may arise from one and the same cause, the disease frequently alternating from one condi‑ tion to the other; and it is particularly worthy of re‑ mark, that very many or most of these cases are either dyspeptics, hypochondriacs, or subjects of *prolapsus uteri,* showing the purely mechanical nature of all the derange‑ ment.

FIG. 5.

FIG. 6.

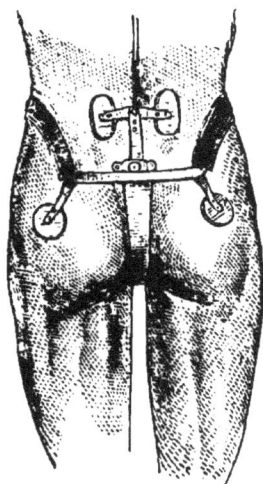

Pregnancy often induces an af‑ fection of the bladder; but cases of incontinence or retention of urine thus produced, will almost invariably disappear at the ap‑ proach of the Pelvic Girdle.

In various cases of affection of the prostate gland, the Body‑ Brace has relieved the tender‑ ness and pain of the irritated part, by simply removing the superincumbent weight.

Fig. 5.—Non-Friction Self-Adjusting Brace-Truss, with Prolapsus Ani Spring.
Fig. 6 represents No. 5 applied.

The *uterus* has five supports, namely, the *cellular tissue* with which it is surrounded, the *vagina*, the *round*, and the *broad ligaments*, and the *peritoneum* or lining membrane of the belly.

These supports are designed to bear the weight of the small intestines, situated below the mesocolon. They also serve as a floor on which the abdominal organs may rest. Their primary object, however, is, to sustain the uterus in the first four or five months of pregnancy, before it has risen above the pelvis, and when it is several times heavier than in its ordinary state; else what would become of this organ? for, surely, if it required all these supports in its quiescent state, there would be an unavoidable falling and dragging, when borne down with weight by the fœtus; a view of the matter, which no rational mind could be found to adopt, as it plainly implies that *prolapsus uteri* is a necessary accompaniment or consequence of gestation. Yet, still it is notorious, that a distinguished member of the medical profession in this city recently asserted on a public occasion, that "*this disease was extremely rare among unmarried ladies.*" And what is this but saying that it is peculiar to those who are *married?* And if peculiar to married life, how far is this from the bare-faced conclusion, that prolapsus is consequent upon gestation?

Now, for the information and benefit of those who are easily led into such preposterous and perilous conclusions, we take the liberty, in passing, to mention, as the result of our own extensive observation and experience in this important matter, that *not one-third* of the cases of gravid uterus complain at all of prolapsus uteri; and also, that a large proportion of these have been repeatedly afflicted in this way, *antecedent to the period of conception.* Should any one, however, after reading this page, still be disposed to the opinion, that "this disease" is *extremely*

BANNING TRUSS & BRACE CO., 704 Broadway, N. Y.

rare among unmarried ladies," we are enabled to state farther, that no less than *three-fourths* of all the cases of *prolapsus uteri* occur to the womb in its ordinary state, the calamity being about *equally* divided between married and virgin life.

But how happens it that this disease prevails to such an alarming extent, when the uterus is not in a gravid state, and therefore not sufficiently weighty to produce any uneasiness; and when, too, it has the advantage of all the strength held in reserve for its enlargement? Why, it is simply because the subjects of this malady, *almost without exception*, are persons in whom the habits of civilized life, favoring general organic derangement and muscular debility of the system, have induced that superincumbent pressure, which, from the very nature of the case, urges the uterus downward, straining its suspensory powers, irritating the nerves of organic life, and establishing the most excruciating pains in the surrounding parts. Indeed, so distressing and terrible is the prolapsus thus induced, that sometimes the uterus is found protruded upon the outer surface of the body!

The effect of muscular relaxation upon the womb and its appendages, is daily producing an almost incredible amount of anguish. Crushed beneath the weight of the fallen organs, as may be seen in Fig. 2, p. 3, the neck of the womb presses with great severity upon the vagina, inducing *leucorrhea*, and many other disagreeable and painful affections. Meanwhile, the *body* or middle portion of the organ is resting either upon the bones or upon the soft parts of the pelvis, causing severe "bearing down pains" in that locality. These pains are often rendered intolerable by the least physical exercise, and are generally attended with a sense of "weight" and "dragging," and oftentimes with a burning pain throughout the system.

The constant pressure which is thus exerted upon the neck of the uterus, brings about a very hardened, irritable, enlarged, painful, and often dangerous condition, which has not unfrequently been supposed to indicate cancer of the womb. We have seldom failed, however, to cure all diseases of the neck of the uterus by the simple application of external support to the abdominal organs, the cause of the irritation being thereby at once removed.

When the *round* ligaments are put upon the stretch, a disagreeable "drawing" or "pulling" sensation will be experienced in the groins where these ligaments are fastened. The *broad* ligaments also, which are spread and fastened over the small of the back, are producing their own peculiar "dragging," "grinding," tormenting pains about the loins, so common to the afflicted female, and so well expressed in that descriptive phrase: "*I feel as though a joint or two of my back were gone.*"

Prolapsus uteri, or falling of the womb, is a disease which has become fearfully prevalent in every community; nor is it confined to married ladies, as has been erroneously supposed by some, whose position, to say the least, should be a sufficient guarantee against all careless observation in medical practice, and whose assertions have received due attention in our preceding remarks.

Only a few, comparatively, of the victims of prolapsus understand the true nature of their difficulty, or know anything about the value of *proper* mechanical support, to say nothing of the *absolute necessity* of its adoption for the eradication of disease mechanically induced. They know not that those who are supposed to have "*recovered*" from this malady, by the internal use of the *pessary*, and the tedious process of lying *perfectly still*, are only *relieved*, it being impossible to *cure* them by any such means; the predisposition to the malady forever remaining as the legitimate offspring of the habitual drooping form, or the fashionable *Grecian bend.*

This dreadful affliction generally comes on by degrees, or, in other words, it *gradually* taxes the subject's powers of endurance; and, if promptly met in its incipient state, by judicious mechanical means, is easily subdued; for, not-

BANNING TRUSS & BRACE CO., 704 Broadway, N. Y.

withstanding the prevalent opinion, that *prolapsus uteri* is not induced by superincumbent pressure, but by primary weakness of the ligaments, we shall hereafter demonstrate (if we have not done so already), that as soon as the *pressure* is removed, the womb will begin to rise by the returning strength of the uterine supports. In order to enable the reader to decide upon the existence of the malady, we give below a succinct description of the same.

Almost simultaneously with the descent of the womb, the back begins to ache, the pains varying in their nature and intensity with the progress of the disease, until they assume the most terrible " pulling," " twisting," " grinding," " wrangling," and " wringing " forms. These pains are usually experienced in the region of the kidneys and in the groins. The back is often represented as being "broken," or "pounded," or "drawn." Severe " *bearing down* " pains, like a ponderous weight at the base of the abdomen, are also felt in the *sacrum*, or rumpbone. The limbs of the patient become heavy and clumsy, and subjected to cramps, numbness, and prickling sensations. The most unremitting and intolerable *dragging* at the breast, which the patient bends forward to avoid—a sense of " *goneness* " at the stomach, swelling of the feet and limbs, constipation and leucorrhea, together with " a feeling as though the hips were loose," are also to be reckoned among the ruinous effects or symptoms of this appalling disease. Although the sufferer may be somewhat lively, and comparatively free from pain in the morning, yet ere noon arrives, she is greatly indisposed, and begins to move carefully, supporting with her hand the lower abdomen, as if she were afraid of jarring or jolting the internal organs ; and not unfrequently does it happen, that before night approaches, her accumulated sufferings will extort cries of anguish and despair.

The above is by no means an exaggeration of the symptoms of *prolapsus uteri*. The *intensity* of the suffering, however, belongs to an advanced stage of the disease. But the *modification* of the symptoms, enabling the subject, as is often the case, though always with more or less suffering, to attend to her ordinary duties, does not at all affect the signs by which the nature of the disease may be discerned. And let it not be forgotten, that in every variety and degree of prolapsus, the safest and most reliable treatment is the immediate application of the Brace and Attachments.

Before proceeding to demonstrate what has been said about the pathology of this disease, by the introduction of cases bearing on the point, we must notice one or two of the objections urged against the use of external support by those whose opinion, in general, is not to be lightly esteemed, and whose errors are therefore the more dangerous. These gentlemen confidently assert, that "prolapsus uteri is *not* caused by a relaxation of the *muscles*, but by a weakness of the *ligaments*; that external support, however concordant with the natural action and bearing of the muscular forces its principles may be, must necessarily press as much *upon* as it lifts from *off* the *uterus*, and therefore can do no good." That such is *not* the fact, is shown by the *invariable* relief given to the sufferer by upward and backward pressure, even when this pressure is made by the mere application of the human hand, that is, by placing one hand upon the lower abdomen, and the other upon the small of the back, lifting and pressing firmly at the same time with the former, in an oblique direction up towards the latter; this peculiar action being the same as that of the muscles which sustain the mechanical relations of the truncal mass.

Again, if mechanical support, imitating the action of the muscles and judiciously applied, will not relieve a case of aggravated prolapsus, we are clearly in a dilemma; for it is admitted that the natural healthful action of the abdominal and dorsal muscles is upward and backward; and that these muscles are the actual organs which support the viscera and preserve their determinate form and position. Of course, then, it follows, that, the more healthy and elastic are these organs, the more perfectly will they effect this perpetual elevation and protection of the pelvic viscera. But, says the objector, "It matters not, how analogous the action and bearing of any mechanical device may be with the action and bearing of the muscles, such device must necessarily produce as much pressure *downward* as upward!" Now, if this were the case we should think it fair to conclude, that, the more active and elastic, and altogether healthy, are the muscles in question, the more will they tend to produce prolapsus, by pressing down a part of the bowels more forcibly upon the uterus. We would also suggest, that such reasoning as this, is not exactly the kind our opponents should adopt, to induce the belief that muscular relaxation may not be a fruitful source of prolapsus and general mal-position of the abdominal and pectoral organs, or that external support, like that supplied by our mechanical devices, will not relieve these effects.

Another very grave objection is based upon the assumption, that even should the support of the Brace, in accordance with our promise, give immediate relief, that relief is not obtained on scientific principles; but in opposition to a known law of the vital economy. It is said, that "if the *natural* function of supporting muscles be *artificially* or *mechanically* performed, their torpor is increased; and that, the natural stimulus on which they depend being thereby superseded, they must soon lose entirely their tone." This is physiologically true; but it is *only so*, when applied to muscles or organs in *health*—as for instance, if the farmer or the blacksmith should quit his vocation, and bandage up his muscles and take no exercise. But *we* come to a *sick* person; to one who is laboring under both local and general muscular relaxation in its most distressing form; where the effects are co-operating with the cause to enfeeble and depress—so much so, that even the most gentle carriage exercise aggravates the symptoms and increases the misery

BANNING TRUSS & BRACE CO. 704 Broadway, N. Y.

of the patient; to one who is totally ineligible, for a time at least, to other than artificial curative means. We come to one, of whom common sense says, "Bind her up, that she fall not to pieces. Hold her comfortably together by such mechanical devices as shall imitate the action and bearing of the muscular forces, and thus maintain the primitive relation of the parts, in order that she may be the sooner enabled to endure such physical exercises as her medical adviser may deem necessary to re-establish the health of the system."

It is also erroneously objected, that those who may recover from prolapsus by the use of the Brace, will be unable, from habit, to lay the instrument aside. And this is deemed by some a sufficient reason for avoiding mechanical support! *Only to think of it!* But after all, there is nothing like putting principles to the test, in searching after truth. Let us therefore make an application of the one before us.

"My friend, you have been for years laboring under great distress, owing to a displacement of some of your organs, and the consequent loss of their vital energy; and you well know that I have tried my utmost to alleviate this distress, by pessaries, tonics, anti-spasmodics, and other hopeful remedies; but all in vain. To be sure, I know of a very simple and comfortable remedy, that would afford you efficient relief, and almost make you forget your troubles. But I must warn you not to think of using this remedy, as there is great danger of relying too much upon it for support; or, in other words, of substituting *the habit of being comfortable, for the habit of enduring pain.*"

To such a warning every sufferer should say "Give me the needed relief, even if the price be the permanent wearing of so comfortable a remedy; let me have the brace." Happily, however, it does not follow that the brace must be worn after cure is effected any more than that the broken leg must go on crutches after the bone is united and strength been restored.

It is also contended by those who oppose the use of judicious mechanical support, that prolapsus is a primary disease, existing in the uterus or its ligaments; that all the attendant affections of the stomach, side, heart, lungs and head, and in fact the whole nervous system, are the result of *sympathy*, and not of mechanical derangement; that inasmuch as prolapsus is a local disease, the remedies should be applied upon the organ itself. Of course, then, it is only necessary to elevate the uterus, so as to give its ligaments time to regain their strength.

This exclusive treatment of the womb, comprising as it does, the simple elevation of the organ itself, involves the persevering use of the *pessary*, a practice which it is humiliating to observe has been adopted and extolled by the loftiest heads and proudest names of the profession; and yet we boldly assert that the pessary, in a clear majority of cases, only makes the case more complete, as it rests on parts that themselves need support, and frequently causes versions and flexions of the uterus.*

It will be remembered that the objection to supporting the viscera by mechanical means, is, that it weakens the powers of life, by *mechanically* performing the functions of supporting muscles. Is it not also claimed for the pessary that it *mechanically* performs the functions of the ligaments, in doing for the uterus what such ligaments ought to do? And will it be said that *this* does not constantly tend to weaken the same, by doing away the necessity for their action? Is not this *especially* the case, if, as is claimed, the original descent of the uterus is *caused and perpetuated by the relaxed condition of these identical ligaments*? But we can readily perceive that the objection applies to the *pessary;* and, with equal force does it apply to almost every kind of mechanical support. We claim for our own mechanical devices, and hereby challenge either private or public investigation, that no such objection can apply; and this challenge we would modestly base upon the *physiological* construction.

* These strictures on the pessary do not apply to judicious intra-pelvic appliances, used in connection with external support, and acting from an external base.

Without stopping to consider, because embraced in our forthcoming review of his pathology of *prolapsus uteri*, the absurd conclusion of the celebrated Dr. Meigs, namely that "a contractile shortening of the vagina, is *the* producing and perpetuating cause of this disease," we would

FIG. 7.

FIG. 8.

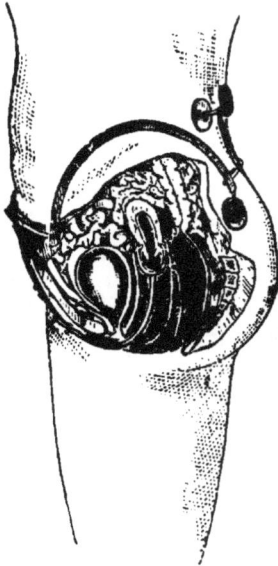

simply ask, What is the natural action of the *pessary* upon that relaxed and irritated state of the vagina, which has been so strangely mistaken for "a contractile shortening," but which we hold to be a remote effect of the relaxation of the muscular bandages of the trunk? Is it

Fig. 7 represents Banning's Improved Abdominal Supporter removing visceral weight and correcting the truncal bearings, while its attachment, Banning's Improved Bifurcated Uterine Elevator, is supporting the cul-de-sac on either side. Thus, while elongating the vagina, restoring the diseased or overtaxed and displaced uterus (without touching it) to its normal position.

Fig. 8 represents Banning's Improved Abdominal and Spinal Shoulder-Brace and Uterine Balance combined, for retroversion and flexions of the uterus.

BANNING TRUSS & BRACE CO., 704 Broadway, N. Y.

BANNING TRUSS & BRACE CO., 704 Broadway, N. Y.

not the perpetuation of such relaxed and irritated condition of the vagina, inducing *fluor albus* and general debility, and otherwise lessening the chances of success, in the subsequent application of other remedial means?

Seeing, then, that the objections urged against judicious external mechanical support, in the treatment of prolapsus, are more applicable to the favorite internal support or *pessary* of our opponents, we now ask the public and the profession, which course of treatment holds out the greatest prospect of relief? In other words, Which course is more likely to restore the normal condition of the uterus and its appendages?

FIG. 9.

FIG. 10.

In cases of flexions and versions of the uterus, we frequently introduce a Banning *Uterine Balance*, but simply to correct the deflection, and as an adjunct to the Brace; this Balance has none of the objections attached to the pessary, as it rests on an external base, and corrects the position of the uterus *after* the visceral weight has been removed by the external brace.

Fig. 9 represents Fig. 8.
Simultaneously poising the superior trunk behind the dorso-lumbar spine; expanding the chest; bracing the weak spine; restoring the normal obliquity of the pelvis, and elevating all the abdominal viscera, from the bladder, rectum, and uterus, the T piece of the uterine balance resting against the posterior cul-de-sac, and compelling the retroverted or flexed uterus to be reposited, producing a lateral contraction of the expanded vagina, and avoiding impingement upon the uterus or rectum.

Fig. 10 combined with Figs. 3 or 4, is adapted to severe cases of prolapsus.

What we would not now speak of, we have already
alluded to, namely, the depression of the mouth or neck
of the womb, into the *vagina ;* by which means the latter
becomes distended and irritated, and its fibres and pores
or ducts relaxed, the secretions being thereby greatly
provoked, and the vessels pouring out a fluid, the morbid
discharge of which is termed *leucorrhœa.*

This malady does its full share in destroying female
health and happiness. No specific internal remedy for
it has hitherto been found, and no age is exempt from its
attack. They who are subject to it, are almost always
afflicted with prolapsus, pain in the back, nervous sus-
ceptibility and lack of energy. In short, they have more
or less of the *drooping* form, and usually complain of most
or all of the effects growing out of the same. These facts
argue an identity of origin and progress in leucorrhea and
prolapsus. But leucorrhea has more commonly been
viewed and treated as a primary disease. Hence it is

Fig. 11.

that in vain almost every variety of treatment has been
adopted for its relief. Of
course we are speaking of this
affection as mechanically induced by general muscular
laxity. Like other *effects*, this
disease is the *cause* of other
difficulties, which is especially
the case, in its aggravation of
prolapsus.

In our professional career,
we found that in making use of the Brace for prolapsus, many were also cured of leucorrhea, the frequency of which occurrence drew our attention to the
subject, and we are now able to say that, with the ex-

Fig. 11, Banning's Improved Straight Uterine Balance, to be combined with No. 3 or 4,
used for anteversion or flexion of the uterus The straight and hollow shaft conceals a
spiral spring. The T piece supports the anterior cul-de-sac.

ception of two or three cases out of the many we have treated, *all* have been essentially, if not permanently relieved by the use of the Brace. And we submit, whether in three-fourths of the cases, the symptoms accompanying prolapsus and leucorrhea' are not the same,—the increase or decline of the one disease being contemporary with that of the other.

EFFECT IN CREATING UTERINE HEMORRHAGE AND PROFUSE MENSTRUATION.

The extent of these diseases would scarcely be credited by any one who has not been brought in contact with their destructive influences. They impair the digestive, nervous, and muscular systems, by deranging the mechanical relations, and then obstinately, yet naturally enough, resisting the *constitutional* overtures of the ablest practitioners, until, at length, their ramified effects lead to the supposition, that the patient is absolutely afflicted with the entire Protean group of maladies, termed female weaknesses.

Almost all of the cases of Uterine Hemorrhage and Profuse Menstruation are caused and perpetuated, to a greater or less extent, by a deranged condition of the truncal organs. And what, indeed, may we *not* expect by a change so great as that from *support* to *suspension*? It is true that some of these cases are much emaciated, and seem not to have abdomen enough to gravitate; and therefore we may not observe in them that morbid shape of the trunk so apparent in our representation of the drooping posture on page 3, where the descent of the viscera may be seen at a glance. But it matters not how inconsiderable may be this descent, so long as there is an actual change throughout the trunk, from support to suspension; for, when the most inconceivably morbid or unnatural change takes place in an arrangement whose susceptibilities are presided over by nerves, and ever so little descent be the consequence, there must always be some modification of the vital functions, either local or general; perhaps both. And when the *uterine* organs are depressed by the abdominal viscera, their functions must either be depressed, exalted, or otherwise modified, by the breach thus made, producing irritation and debility in the nerves of organic life, by continually pressing upon them, and stretching their connections with the surrounding parts.

These diseases are always benefited by the application of the Brace. It matters not whether they have a local, constitutional, or mechanical origin. But when they exist as the legitimate effects of muscular weakness, their speedy removal by the use of this instrument, is no longer a matter of conjecture.

EFFECTS OF PRESSURE IN PREGNANCY—THE PELVIC GIRDLE.

The period of pregnancy is beset with a full share of troubles. Among these are pain and weakness of the hips and limbs, occasioned by pressure on the sensitive and ligamentous tissues; also great weakness and pain in the back, together with a darting sensation up the spine into the head, inducing confusion of mind, dizziness, ringing in the ears, and "strange feelings;" cramps in the muscles of the abdomen; weight, distension, and costiveness; urinary incontinence, or retention; varicose veins, and swelling of the limbs. Here, indeed, is a startling combination,—a sort of digest of human suffering. And is it any wonder that all this should produce abortion? Is it any wonder that the life of the sufferer should be worn out before delivery; or that her offspring should so often drop still-born into the world? Now, we hold that almost every pang peculiar to the period under consideration, is *mechanically induced by distension, weight, and pressure*, and that, therefore, instead of depending as hitherto upon cathartics, the lancet, and paregoric, for temporary relief, these distressing complaints, which have rendered

BANNING TRUSS & BRACE CO., 704 Broadway, N. Y.

this portion of female life a terror, may be entirely removed by such judicious mechanical support, as the girdle in question is every way fitted to impart.

And here we cannot but bespeak in our own behalf, an inspection of this mechanical device, by our medical brethren; for, surely, every member of the profession must needs welcome an auxiliary of the kind,—one which, as soon as any of the above obstinate symptoms shall prove themselves to be *non-cognizant of medicine*, is abundantly able to step forth and make a *mechanical* adjustment of the matter. For our own part, we trust we are duly thankful for the light we now enjoy, and that we shall ever be ready to make our humble acknowledgments for each additional ray. It is gloomy enough to reflect on the amount of loss and suffering which the want of knowledge upon the subject is daily entailing upon thousands; and, therefore, it is with no ordinary feelings that we have embarked in the dissemination of our views of mechanical support. Thousands of abortions, still-born children, bad labors, and deaths, might be prevented by the timely application of the Pelvic Girdle.

EFFECT IN THE PRODUCTION OF VENTRAL, INGUINAL, AND UMBILICAL HERNIA OR RUPTURE OF THE THIGH, GROIN AND NAVEL.

On each side of the lower abdominal walls are two openings through which pass certain cords. These apertures are liable to relaxation or distension from various causes, the bowel often protruding like a sac, producing a hernia or rupture; whilst this malady has many degrees of severity, it *always* renders life comfortless, and often proves *fatal*. In the relaxed or unsupported state, the *pressure* of the entire visceral mass against the inguinal rings is abundant cause for hernia, though it is frequently caused by running, lifting, coughing, and even by a mistep; and if the encroachment of the internal organs upon the rings

distend them and thus effect protrusions (for how can
a passive bowel resist the pressure which is forcing it
through the aperture?), *the protection of the rings against
further encroachment, by the removal of the superincumbent
weight,* is certainly the common sense method of curing the
rupture. This can only be accomplished by applying support to the abdominal organs. To this end our appliances are admirably adapted; and if all feeble persons
of lax habits, who are troubled with "prickling" pains
and weakness in the lower abdomen, should adopt the
precautionary measure of wearing them, many of the
most disheartening cases of rupture would be avoided.

There are but few contingencies more common and
dangerous than rupture, it being estimated that at least
one in seven of adults is so afflicted, and yet it is a melancholy fact, that with all the inventive genius of this age,
hitherto there has not been produced a hernial truss
based upon proper principles, the trusses heretofore
made being clumsy contrivances, without the least lifting
power; the inventors or manufacturers thereof being
apparently ignorant of the laws of anatomy and physiology. In all their trusses the principal object they have
appeared to keep in view has been to "plug up" the
Hernia by strong pressure upon the external inguinal ring,
forgetting that by so doing they only caused the breach
to be more extensive. Among the defects of the ordinary
truss of to-day we would mention :

1st. Total absence of any power to remove the prox-
imate cause, viz., visceral weight.

2d. Owing to absence of lifting power, greater pressure
upon the external ring.

3d. A pad so large, with pressure so great as injuri-
ously to compress the spermatic cord. For, in addition to the absence of the lifting power, the large pad
covering so much of the wall, calls for greater pressure to
retain the Hernia. Who, for instance, would attempt to
stop the leak from a gimlet hole by placing over it the

BANNING TRUSS & BRACE CO., 704 Broadway, N. Y.

entire hand? If they did they would have to use great force to accomplish it; but by putting the point of the finger over the aperture, a slight pressure would prevent the leak. So with Hernia, the smaller the space covered by the ball the more effective it becomes. Patients often remark, on seeing our Hernial apparatus: "Oh! Doctor, *that* wont do, because I have an *awful, awful* big Hernia;" and they will take off the appliance they have been

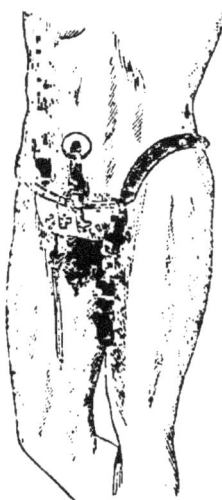

Fig. 12.

wearing, which will look as though large enough for two or three ruptured men. They forget what the farmer boy said, viz.: "That a very large drove of cattle can come through a very small gate."

4*th.* Without any exception, all the trusses heretofore have pressed upon the back. This they can not do without injuring the sacral (spinal) nerves, and frequently paralysis results therefrom.

5*th.* The cushioned pads inevitably cause relaxation, which is a very great defect.

6*th. All* other umbilical trusses *crowd down* the bowels, they procuring their retentive power by encircling the waist; while ours act from below.

7*th.* Their friction and liability to displacement complete the category of our complaints against them.

This essay has not borne thus hard upon trusses, their inventors and manufacturers, through any feeling of *rivalry,* but *because* the defects it has pointed out *do* exist, and have caused an immense amount of human suffering

Fig. 12 represents the Banning Non-Friction Self-Adjusting Brace-Truss, curing double inguinal, femoral, and umbilical Hernia. It acts upon the principle of elevating visceral weight from the Hernial openings, and gives but slight pressure, not on the Ring, but in the canal between the two rings, causing a consequent contraction of them, and in sixty per cent. of the cases a speedy radical cure. It can not become displaced. Is light, cool, self-adjustable, and an absolutely "Non-Friction Truss."

and it seems strange that the profession should have left the inventing of proper Hernial apparatus so much to mechanics who possessing no knowledge of physiology, or to charlatans, who, taking a piece of hoop-iron, construct a truss, and announcing "Eureka," offer it to the public—not only to the damage of the buyer's purse, but, what is of much more importance, to the serious detriment of his health. For illustrative cases and description of Banning's Hernial appliance, see pages 61, 62; Figs. 5 and 12.

Effects of Dragging upon the Parts Above.

MORBID EFFECTS OF RELAXATION UPON THE VOCAL ORGANS. (BRONCHITIS OF PUBLIC SPEAKERS AND SINGERS—THROAT AIL—WEAK AND UNNATURAL VOICE.)

The possession of a clear, smooth, and melodious voice, depends upon the correct action of the abdominal muscles, which lie at the foundation of vocal philosophy, and are admirably adapted to force the well-packed internal organs against the lungs, thus driving the air in a direct line along the natural course of the slightly curved vocal tube.

Now that the body is drooped, the shoulders and chin depressed, and the internal organs deprived of support, how can the delicate texture of the vocal tube resist that concussion and abrasion which must inevitably be produced by this condition? For it is plain that the application of the expulsory power must now be made either obliquely or at right angles. In other words, by sinking or drooping the shoulders, and contracting or depressing the chest, the power must be applied at opposing points, which will, of course, produce a general concussion in the cells and bronchia, and cause the air so to press downward upon the diaphragm, that if a vocal apparatus were located at this point, we should doubtless **have** its utterances in no uncertain tone to support our position.

The air being thus scattered in all directions, a portion of it goes oscillating and rasping up the throat in confused and ragged currents, until it strikes the roof of the mouth, when, by the acute angle in the passage, it is again concussed and scattered, a part of it seeking egress through the nose, that being the straightest and most convenient exit, whilst the whole operation is accompanied with a flat, jarring, and unpleasant voice. Another effect is, that the air-tube being curved, and the outlet contracted, the air moves in a ragged stream, in lieu of a pure smooth current, injuring the bronchial or lining membrane of the throat and windpipe, and of other smaller tubes, mechanically irritating their surfaces, and inducing at the same time a sense of dryness and aching, especially in speaking and singing.

The surface is soon exposed, pimples appear, and the result is, a genuine specimen of bronchitis, or, as it is often erroneously termed, "the throat disease of public speakers." These views are strongly corroborated by the fact, that almost all broken-down public speakers and singers complain much of weakness and goneness at the stomach, and pain in the back. We also find that the subjects of Bronchitis, or Throat-ail, are the relaxed and feeble; those who generally sit, stand, or walk in the curved or drooping posture, with the head bent forward, the chest compressed, and the abdomen relaxed; who are deprived of regular exercise in the open air. By far the greatest number of cases are found amongst mechanics, school-teachers, and ministers of the Gospel, who preach *calmly* by note from a low desk.

MORBID EFFECTS OF RELAXATION IN PRODUCING AFFECTIONS OF THE LUNGS.

The diaphragm, or floor of the lungs, being now deprived of its usual support, by the withdrawal of the abdominal organs from their natural position, its fibres can no longer be put upon the stretch; in consequence of which, they soon lose their power to contract, and are accordingly dragged upon by the falling organs. This inevitable result of a relaxation of the abdominal belts, and consequent falling of the bowels, causes, in conjunction with the influence of the surrounding atmosphere, a compression of the waist, and an elongation of the chest, as well as severe depression at the stomach; and, as the body is also of the drooping form, the weight of the trunk prevents the expansion of the short ribs, by crushing in the chest at its most movable part.

The restriction thus imposed upon these ribs, destroys the power of the intercostal muscles in this part of the chest, to say nothing about the separation of their front extremities. Inspiration must therefore be chiefly effected through the instrumentality of the superior intercostals; and hence it is that in all cases where the stomach is retracted, and the waist compressed, inspiratory efforts will not expand either the sides or abdomen. Respiration, accordingly, becomes short and quick, being performed, as in the case of consumptives, who are troubled with shortness of breath, by a heaving of the upper portion of the chest, with the aid of the lobes in that locality.

The philosophy and mechanism of breathing being thus interfered with, the most dreadful effects are entailed upon the sufferer, such as bleeding at the lungs, cough, shortness and tightness of breath, as if breathing through a sieve, dead, dull, fixed, or wandering pains in the breast, together with a sense of pulling in the centre of the chest, or drawing at the pit of the stomach, as though there was something in that locality resisting all

BANNING TRUSS & BRACE CO., 704 Broadway, N. Y.

BANNING TRUSS & BRACE CO., 704 Broadway, N.Y.

efforts to inhale a full breath, and defeating every attempt to sit erect and develop the chest. This pulling sensation is explained by the fact, that when the diaphragm is dragged down from its elevated position, as is now the case, the mediastinum, or partition between the right and left lungs, is put upon the stretch, and thus placed in a very unnatural situation, it being no part of its office to sustain the organ to which it is attached. It is this stretching of the mediastinum that makes all consumptive persons complain of a sense of pulling, or of tightness or stricture in the centre of the chest, the tense membrane becoming relaxed, and the sensations relieved whenever the diaphragm rests upon, or is firmly supported by, the organs below. This deranged condition must permanently compress the lower, and throw a great burden upon the upper lungs, thus increasing the labor of respiration, and thereby establishing sundry local and constitutional difficulties, *the mechanical origin of which* accounts for the fact, that whilst they are always relieved, and often eradicated by the judicious elevation of the visceral mass, and the consequent shortening and expanding of the chest, they *never* yield to exclusive internal treatment.

Affections of the Lungs, though varying greatly in degrees of fatality, are at all times justly regarded with feelings of apprehension, for it cannot be expected that such delicate organs as the lungs can long remain uncured, without establishing tendencies which too often result in the termination of life.

Whenever the usual symptoms of Consumption exist, the pathology of the disease is the first question to be settled, the prospect of recovery being materially affected thereby. Is the difficulty of an organic nature, or does it arise from simple *mechanical* derangement? For, when the substance or structure of the lungs is invaded, the prospect of cure is almost in exact proportion to the extent of the disorganization.

We now proceed upon the supposition, that morbid relations are existing in connection with the above-named symptoms of disease ; and where these do not exist, then, although all the symptoms are complete, our views do not apply to such cases. They must be treated in the best possible way, with other than physical or mechanical remedies. But such cases do not often exist ; for, however certain it may be in any particular case, that this dreadful malady has not been caused by a morbid relaxation of the muscular bands of the system, it will almost invariably be manifest, that it has ravaged and deranged the machinery of the trunk to such an extent, that no treatment whatever can impart essential relief, without the *aid* of mechanical support.

The general mechanical physiology of man in a healthy state, requires that the body should be erect, the shoulders high, the chest full, the short ribs playing freely, the lungs at perfect liberty, the abdomen round and firm, the stomach full, the back well curved inward, and the bowels elevated and supporting the diaphragm, as in Figure 1 and page 3. Whereas, in almost every individual complaining of the above symptoms, we find the shoulders drooped and brought forward, the back strongly curved, and the whole trunk sunken and relaxed, the breast bone being forced inward, and the ribs and lungs compressed.

The bowels having fallen, and the abdomen being hard and large at its base, and small and soft at the waist where it is much retracted, the stomach, liver, and spleen, and other superior organs, having lost their support, must accordingly descend ; so that, in the standing posture, particularly, the patient will complain much of weakness of the back, weight in the abdomen, sinking at the stomach and tightness in the centre of the chest ; and, indeed, whenever he attempts to straighten up, draw back his shoulders, and throw out his chest, he immediately relapses, saying that it pains him to stand erect ; that it pulls at the pit of the stomach, increases the tightness in the breast, and shortens his breath ; all of which is accounted for by the fact, that, in raising himself he is also obliged to raise the bowels, the tension and gasping to which we have alluded being thereby produced.

This deplorable state of things can only be relieved by the due expansion of the chest and lungs. And how is this to be accomplished? By internal and constitutional remedies, or by judicious *mechanical* means? Evidently by the latter, as *medicine* can produce no change in the *form*, or remedy physical defects. First, the abdomen must be properly elevated, so as to lift the superincumbent mass upward and backward, by means of which the lungs will be supported, the chest enlarged at its lower region, and the tightness and stricture removed,— the author's Body-Brace, and abdominal and spinal shoulder Brace, as we shall presently show, being the only instruments adapted for this purpose. The patient can then stand erect, or place himself in any required position, without the usual pain and stricture in the breast and stomach, the diaphragm also being thereby elevated and made tense, and caused to contract more perfectly in respiration.

MORBID EFFECTS OF RELAXATION UPON THE HEART AND BRAIN. (PALPITATION OR FLUTTERING, SINKING OR FALLING OF THE HEART, ; FULNESS, TIGHTNESS, AND DIZZINESS OF THE HEAD ; CONFUSION OF IDEAS, AND LOSS OF MEMORY.)

It will be recollected that the heart is retained in its natural state by ligaments, arteries, and veins, as well as by the diaphragm, from which it derives its chief support. But now that the diaphragm is dragged down from its proper position, the heart also, with which it is firmly connected at its base, must necessarily descend, oppressing and irritating the nerves of organic life that preside over its involuntary action, and producing every form of palpitation, or " fluttering," as well as " sinking " or "falling" of the heart. Indeed it often happens that this increased action, or beating of the heart, is found sufficiently low down to indicate a positive change in its locality. Note, too, with what hasty and firm strokes the human heart beats at the invasion of individual rights, making the tired muscle ache with the exertion of its giant force, as it sends the blood to the very surface of the body. See also how at times it leaps for joy. Behold it, on the other hand, in sorrowful and almost pulseless depression, and then say, whether, if such comparatively trifling *external* circumstances may, through the mind, produce such modifications in its action, *internal* circumstances of a mechanical nature, bearing directly on the heart and putting all its fastenings or moorings upon the stretch, may not be expected to superinduce very important modifications in its *position*.

When palpitation of the heart is once originated, no matter how, a multitude of effects or results will naturally follow. Prominent among these is an increase in the size and strength of the organ. This morbid condition of the heart is caused by the agitation of its fibres, and may be illustrated by the fact, that the limbs most used by the farmer, blacksmith and dancer, become, from the natural stimulus of exercise, very large and strong. This enlargement, though but a mere effect or symptom, will eventually become an active and perpetuating cause of disease.

t may be well to remark in this place, that whenever _pitation of the heart is produced by an inflammatory affection of the spine, which may at all times be ascertained by smart pressure on the spinal column, the ordinary routine of treatment, adopted by the profession, should at once be resorted to, though such treatment should not be too long continued. If the relief given is not radical, or likely to prove so in a very short time, it may be proper to use other means; for, partial relief is not always to be regarded as evidence that the further prosecution of the treatment is advisable. It may not

BANNING TRUSS & BRACE CO., 704 Broadway, N. Y.

be possible for such treatment to do more than relieve the acute and aggravating symptoms. The remaining effects may be chronic ; and they may also require that the body should be supported by the elevation of the abdominal organs.

Another serious difficulty arising from palpitation of the heart, has its bearing upon the lungs. The irregular and unnatural action of this state, may induce a derangement of the valves which admit the passage of the blood from one room of the heart to another. For instance, the arteries send the blood to the lungs from the right side of the heart, and the pulmonary veins carry it back to the left side, from whence it is sent over the whole system. Now, if the valve that defends the passage of the blood from the first to the second room in the left side of the heart does not act, or becomes bony or gristly, of course the blood will be obstructed in its passage from the lungs through the heart, and so on throughout the system. The blood will, therefore, accumulate in the pulmonary veins, and becoming gorged in the lungs, will inflame and produce in them a sense of suffocation, the complexion of the patient being blue, the lips livid, and the face flushed. To free themselves from this feverish state, the lungs secrete a thick, tough mucus, which, together with a troublesome cough, leads the patient to conclude that the difficulty is a primary disease of the lungs. Hence the importance of being able to determine at the outset, whether the disease is symptomatic,—*the mere effect of a remote mechanical cause, and calling only for mechanical aid.*

The vessels of the heart divide into those which run into the head and arms, and those which supply the lower trunk and extremities. The distance from the starting point to the head being shorter than the other, an increase of action in the heart must send an unusual quantity of blood to the brain ; and, as the brain at all times so completely fills the cavity of the skull, that even the courses of the blood-vessels are imprinted upon it, this surcharge of blood must be attended with painful and injurious results. The cranium being formed of bone, and therefore inelastic, and the blood being thus forcibly ejected into the delicate organs that already fill the cavity, of course the nerves of seeing, hearing, tasting, smelling, and feeling, must undergo severe compresssion. A sense of fulness and tightness will also be experienced in the head. Persons of nervous temperament will in cases like these find a recumbent posture uniformly attended with headache.

Such patients will often, after reaching or stooping, carrying a weight, or ascending a hill or flight of stairs, suddenly complain of dizziness, blindness, or confused vision, as well as unnatural and frightful objects. The patient will reel and stagger, and seize hold upon the nearest object for support, complaining at the same time of a sensation of water in the head, and of ringing in the ears, together with a confusion of ideas and loss of memory, the attitude being fixed, and the head firmly clasped with both hands. When these symptoms have passed over, the patient will speak of having felt a creeping sensation travelling up the spine, entering the brain, and spreading out in all directions. Sometimes this affection passes off quietly, the patient moving gently and looking around as if quite surprised at finding himself alive; at other times, it is immediately succeeded by bursts of tears and sobs, the patient not being able to assign any reason therefor. On some occasions the patient will scream, seem delirious, and talk incoherently. Persons thus afflicted will be often telling that life is a burden, and that they have no comfort. They are ever looking for death, and yet are filled with terror at its approach.

From the foregoing facts, we may learn that the probable cause of nervous troubles is a general or local muscular relaxation, not an affection of the truncal organs; and, also, that such diseases do not exist in the imagination or fancy, nor depend upon a strong or weak mind.

They cannot be controlled by the most powerful intellectual influences. We might as well talk tactics to the wind, as expect the mind to curb or govern nervous affections brought on by mechanical causes. What then shall we say to those who address such cruel words as these to the sufferer: "O, you are only hysterical! You will get better by and by;" unmindful of the fact, that when human nature is thus bowed down, it more especially needs the kind look, the sympathizing voice, and the supporting arm.

MORBID EFFECTS OF RELAXATION UPON THE STOMACH, LIVER, AND SPLEEN. (DYSPEPSIA—LIVER COMPLAINT—CHRONIC INFLAMMATION OF THE SPLEEN.)

The *dangling* condition of the stomach, in the unnatural state, also enables us to account for those peculiar sensations of "vacancy" and "goneness," so common to the dyspeptic. And hence it is, that persons afflicted with dyspepsia are so often compelled to lie down immediately after eating; the relief afforded by the recumbent posture, under such circumstances, being obvious to every one, and quite enough to throw considerable light upon that mysterious, yet oft-repeated declaration of those who have *graduated*, namely, that they are thus enabled to "*enjoy* a sort of *comfortable misery.*"

The fact should not be overlooked, that persons afflicted with dyspepsia may complain either of a voracious appetite or of no appetite at all. Still, *whatever* is eaten lies heavily on the stomach, producing a sense of "flatness," or "weight," or "load," or "burning pain," which is often insufferable. These unpleasant sensations are accompanied with a belching of gas, or the throwing off of a hot acid fluid, that seems almost to skin the throat and melt the teeth of the invalid.

Seeing, then, the legitimate results of a relaxation of the abdominal muscles, or, in other words, of the habitual deviation from the erect posture, and the consequent changes in the mechanical relations of the vital organs, let us seriously reflect whether dyspepsia, or other maladies thus provoked, can be eradicated by other than mechanical means?

Dyspepsia may indeed result from causes requiring constitutional treatment. But when it arises from a *mechanical* cause, which is almost invariably the case, as shown by the fact that every confirmed dyspeptic is of the *drooping* form, is it *possible*, we ask, for internal remedies to effect a cure? For cases see page —.

The Liver.—The liver is pressed upon in all parts by the ribs, lungs, and bowels. Its function is to receive the returning blood, from which it draws out or secretes an element called bile. This secretion, by virtue of surrounding pressure, which it relies upon instead of muscular tissue, is thrown into the first bowel to assist in digestion. But now that the stimulus of pressure is removed by morbid relaxation, the liver ceases to perform its natural functions; and, being very heavy, it must, in its present unsupported condition, make large demands upon the ligaments connecting it with the diaphragm; which, by such unnatural pulling, becomes almost inverted. This dragging upon the diaphragm causes the sufferer to sit or stand with his hand upon his right side. It also disposes him to lean in that direction.

The effect upon the system, of so great a change in the mechanical relations of this organ—a change from *support* to *suspension*, has induced the bestowal of a fearful amount of medication upon so-called diseases of the liver. We would not, however, be understood to favor the view that these mechanical difficulties may not exist long enough to induce other symptoms, resulting in the establishment of a compound ailment, requiring *both constitutional and mechanical treatment.* But even in such cases, common sense teaches us that the *chief* reliance for a radical cure, is upon *mechanical* aid, that being best adapted to remove the *cause* of the complicated ailment.

Invalids should be urgently pressed for a description of their feelings. The *language of sensation* is of infinite importance in determining the nature of chronic disease. Take, for instance, the following by way of illustration. After several unsuccessful attempts to ascertain the peculiar sensations of a distinguished lady in a case of supposed *liver complaint*, she replied : "Why, Doctor, it feels just as though it was hanging from where it is hitched!" This was certainly quaint, but it told the story, and enabled us promptly to decide that the affection was purely mechanical.

The Spleen.—Next in the order of remark comes the spleen. It will be remembered that it is situated in the left side, under the short ribs. Its displacement, as well as that of its suspended associates, causes a tension of the ligaments connecting it with the diaphragm. This organ is accordingly much disturbed, being thereby placed between two pulling forces—a position eminently calculated to produce chronic or acute inflammation in the several organs concerned. The spleen is the seat of many of those chronic difficulties which have been so long and unsuccessfully treated by the profession as primary diseases of the organ, under the name of *affections of the spleen.* The more common of these affections are lameness and tenderness, with a "dead," "dull," "deep" and constant pain in the left side, for which cuppings, leechings, issues, and every other conceivable counter-irritation have been unsuccessfully applied. But let us lift the abdomen of the patient and press on the small of the back, thus creating the natural state, and the patient will at once find relief. What more positive proof can we have of the need in such cases of *the Brace?*

MORBID EFFECTS OF RELAXATION IN THE PRODUCTION OF CHRONIC DIARRHŒA AND DYSENTERY, AND CHRONIC PERITONITIS, OR GENERAL TENDERNESS OF THE ABDOMEN.

The bowels now lie inactive upon the front bone at the base of the abdomen, below the influence of muscular contraction, causing a sense of " deadness " and "weight " to be experienced in that locality; and, if this descent does not pull down the stomach, liver and spleen, it does, to say the least, leave them to the influence of their specific gravity, in which condition they must necessarily produce a pulling or dragging upon the diaphragm and other organs above, affecting more or less the digestive powers, and destroying the tone of the bowels. Such a state of things is well calculated to result in diarrhœa or dysentery, which, when thus mechanically induced, will undoubtedly resist the skill of the practitioner, and gradually assume a most distressing chronic form.

Of course there are many things which affect the bowels and suddenly produce diarrhœa and dysentery, without reference to any displacement of the organs. But if these affections are not checked, they also must assume a chronic form ; for, as soon as the bowels are diminished in volume by the frequency of the discharge, they will of course recede from the abdominal walls, and, being thus deprived of their usual support, must fall to the bottom of the pelvic cavity, thereby removing the stimulant or tonic influence of the muscles pressing on the viscera and producing the same sensation as when these diseases are induced by muscular laxity. Indeed, we might as well, after all, consider this form, also, of chronic diarrhœa and dysentery as having a *mechanical* origin; for, although it does not *arise* from mechanical derangement, yet nevertheless, it clearly appears that its duration actually induces muscular debility. For cases see page

BANNING TRUSS & BRACE CO., 704 Broadway, N. Y.

MORBID EFFECTS OF RELAXATION IN THE PRODUCTION OF SPINAL AFFECTIONS AND DISTORTIONS OF THE HIPS AND SHOULDERS.

The bowels having moved forward, out of the axis of the body, are now producing a leverage on the spine; and, notwithstanding the powerful antagonistic efforts of the dorsal muscles to resist this leverage, the chest, shoulders and head will be brought forward, the lungs compressed, and the abdomen enlarged.

The reader will remember that the body, when *erect*, rests on the *processes*, or central portion of the spinal column, as in Fig. 1, and that when it habitually bends forward, as in Fig. 2, the pressure is transferred to the body of the bone itself, the elastic gristle or cartilage between the bones not being sufficient to protect them from the injurious results of so gross a violation of the laws of health. This pressure being constant and unnatural, will, if long continued, be followed by great tenderness and absorption. The rough surfaces of the bones will thus be brought together, when severe spinal irritation, ulceration, and ruinous spinal curvature will be the inevitable result.

The fatigue and exhaustion consequent upon the subjection of the muscles to such constant and extraordinary exertion, may be compared to that which we experience in ascending a hill. At first our limbs are able to double their exertions, but they soon flag; and as the top is neared, we are required to put forth the most powerful efforts to make even a small advance. When thus weakened, if compelled to struggle for a great length of time, they will become tender, tremulous, and unable to perform the simplest task. So with the muscles of the back. Strong and successful as they prove themselves in long continued efforts to keep the body erect, they must, like every other part of the human system, eventually yield to the destructive influences of a relaxation of the abdominal muscles.

BANNING TRUSS & BRACE CO., 704 Broadway, N. Y.

It will not be denied that spinal curvatures and distortions of the hips and shoulders have become alarmingly prevalent. It is no uncommon thing to see spines curved like an Indian's bow and the letter S, and others bulging out backwards or forwards, whilst irregularities of the shoulders and protrusions of the hips, present themselves at every step. These distortions are most numerous among females, so much so, indeed, that they are beginning to be estimated among diseases peculiar to the sex.

Dr. Warren, of Boston, in his valuable little treatise on physical education and the preservation of health, says, "I feel warranted in asserting, that, of the well-educated females, within my sphere of experience, about *one-half are affected with some degree of distortion of the spine!*" He also quotes a popular foreign writer in support of this assertion, who, speaking of the lateral curvature of the spine, says, "It is so common, that out of twenty young girls, who have attained the age of fifteen years, there are not *two* who do not present very manifest traces of it."

The same writer correctly observes, that, "Causes which affect the health and produce general weakness operate powerfully in producing affections of the spine, in consequence of the complexity of its structure and the great burden it supports. When weakened, it gradually yields under its weight, becomes bent and distorted, losing its natural curves, and acquiring others, in such directions as the operation of external causes tends to give it ; and these curves will be proportioned, in their degree and in their permanence, to the producing causes. If the supporting part is removed from its true position, the parts supported necessarily follow, and thus a distortion of the spine effects a distortion of the trunk of the body."

The prevalence of spinal affections among females, is owing somewhat to the fact, that ladies sit longer at the drawing-table, and live more upon romance than upon exercise and fresh air. But the chief responsibility rests upon the present system of education, which discourages in young ladies the development of muscular strength, and teaches them to look no one in the face, but rather to observe that perfect caricature of human dignity and symmetry, the Grecian bend; and until parents see the importance of educating the *bodies* as well as the minds of their children, physical weakness and spinal deformity must necessarily abound.

The drooping or lounging attitudes which young ladies are thus allowed or instructed to assume, must, of course, relax the muscles of the abdomen, put those of the back upon the stretch, and bring more pressure upon one hip than upon the other; so that, finally, to prevent falling, the individual leans towards the hip, where there is the greatest pressure. Of course, then, the axial line cannot run down the spine, but must cross the same; and if the person is sitting upon the right hip, it must run from this point to the left shoulder, passing diagonally through the body. This unusually great and peculiar pressure upon the spine will produce in it a curve between the shoulders, pushing one of them outward, making it seem larger than the other, and also causing it to appear diseased, when it only wants to be put in its proper place. (See Figs. 13 and 15.)

The muscular power being now unequal, and pulling unduly upon one side, the other shoulder must of course become depressed. The manner in which the antagonizing power is lost may be illustrated by standing or resting upon one foot. It will then be perceived that the muscles of the side on which we stand are unusually active; and that in order to preserve the body in its new axis, they draw down the corresponding shoulder. Of course the muscles of the other side become relaxed, whilst the shoulder in turn is elevated above the other. This inequality, or elevation, this "growing out" of the shoulder, as it is often called, which is always accompanied by a single or double curvature of the spine, is

BANNING TRUSS & BRACE CO., 704 Broadway, N. Y.

more common than any other spinal affection, and is often found without any visible forward drooping of the shoulders.

FIG. 13.

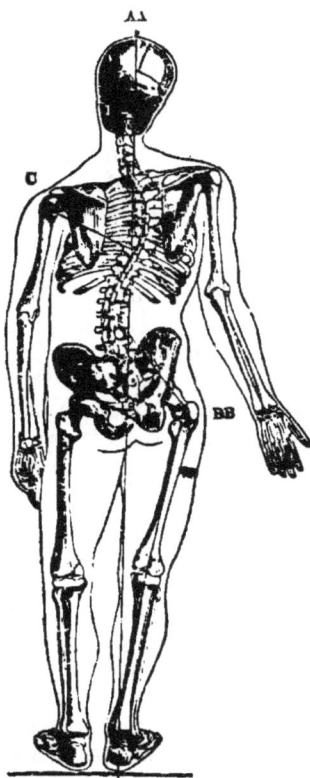

FIG. 14.

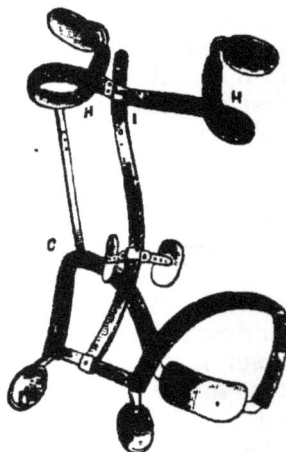

It will be seen by referring to our side views of the erect posture, Figs. 1 and 2, that the axial or middle plumb line, from the crown of the head to the feet, passes through the spinal marrow at the nape of the neck, and in the hollow of the back; that it touches the spine only at these points. These points, therefore, must all be in line when the body is erect. The *head* only is allowed to move out of line, and not the shoulders.

Fig. 13 represents the body supported mainly on right foot. A A, perpendicular line, from centre of head to right heel, showing the head to be still vertical to the basal point. B B, angular line, indicating the direction of gravity against the lumbar spine and showing it to one side. C, line showing the weight of the head and left shoulder to be in the interest of a dorsal curve to the right.

Fig. 14 represents the Centripetal Spinal Lever, for Double Spinal Curvatures. See Figs. 15 and 16.

BANNING TRUSS & BRACE CO., 704 Broadway, N. Y.

FIG. 15.

FIG. 16.

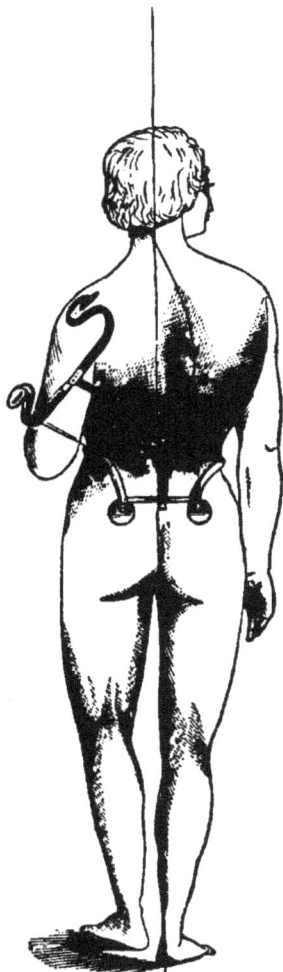

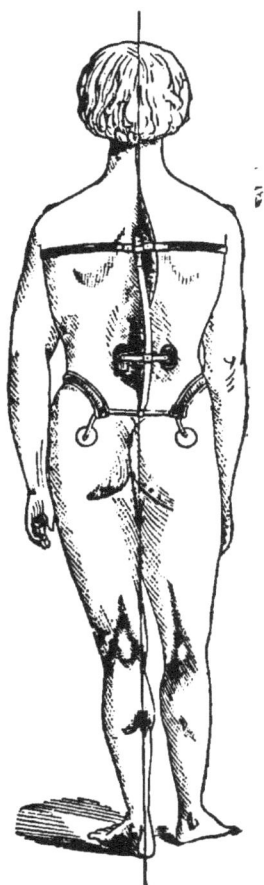

Fig. 15 represents the Centripetal Spinal Lever accomplishing nothing, its lever powers not being brought into activity by being brought around the shoulders.

Fig. 16 represents the Centripetal Spinal Lever in full activity, elevating and drawing out the left shoulder; supporting the lumbar curve to the right, and aggressively restoring the body to its axis, and so crushing out the curvature by means of the very gravity which caused it.

Their drooping depends on a previous movement at the next point of motion below the nape of the neck, which is in the hollow of the back, at D. So, also, when we bend forward or backward, it is through the agency of this part of the spinal column. This being the case, it follows that all deformity and misshape about the chest and shoulders, where there is no tuberculous or other disorganization of the spinal column, must be referred to the *hips* and *small of the back*.

It should be remembered, therefore, that when we stoop or lounge, the *chest* does not advance or fall, as it seems to the eye; the false impression in such cases being produced by the simple fact that the point of motion, or fulcrum of the spine, recedes from beneath the chest. In other words, the middle of the body gets behind the shoulders, as clearly demonstrated by the circumstance, that in both the natural and drooping figures, the perpendicular lines touch the same points about the head and feet. So that in order to straighten the drooping form on page 3 or to effectually cure round shoulders, it is only necessary to push the middle of the spine at D against the *axial* line B B, leaving the shoulders entirely undisturbed. This brings the portion of the spine which had receded, namely the *middle*, directly under the shoulders, as represented by the mathematical lines in Fig. 1.

For example: Stand perfectly erect. Then let a friend place his thumb and forefinger against the spinal column, in the hollow of the back at D, and press firmly enough to resist the weight of your body. You will find it absolutely impossible to droop your shoulders in the least, or to bend your spine at any point. You may move your *head* forward, but *bow* you cannot, and, if you persevere in this effort, your whole body will fall like a straight stake, over the ends of your toes. Your friend will then tell you that in your efforts to bend your body, the hollow of your spine pressed backward with great force against his thumb and finger.

The reason why there can be no natural and voluntary

BANNING TRUSS & BRACE CO., 704 Broadway, N. Y.

or involuntary motions of the shoulders, when the hollow of the back is held in axis, is that this point must always make an *antecedent* retreat. This is evident from the fact, that if your friend had held his fingers an inch from the spine at D, you would not only have been able to bow with ease, but your spine would have retreated to his thumb and finger, and the more you drooped your shoulders or bent your body, the harder would your back have pressed against his thumb and finger.

FIG. 17.   FIG. 18.

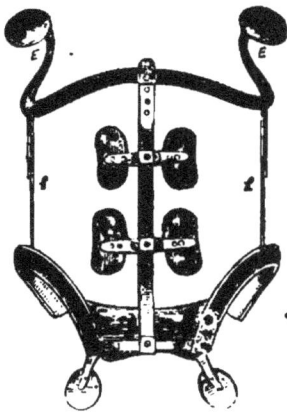   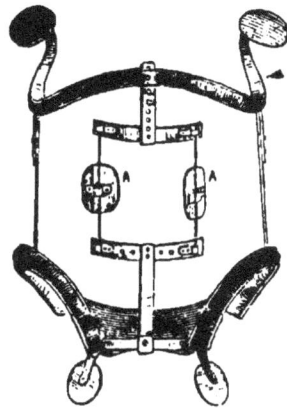

To render the argument still more conclusive, suppose you assume the drooping posture. Your friend will then see that the hips and hollow of the back are not in range with the centre line B, but with the third and rear line. Let him now push you firmly at D, and you will either fall upon your face, or find your back, hips, abdomen and stomach all coming forward, the hollow of the back, and

Fig. 17, the Spinal Prop, consists of Fig. 4, with the addition of ff, which are side-posts resting upon the arches of Fig. 4, and converting E E into unyielding crutches.

By this the viscera are elevated, the shoulders drawn back, and the proper forward spring of the lumbar spine preserved, and a tender or carious spine is relieved of pressure. The subject can exercise freely, and threatened disorganization and curvature are averted.

Fig. 18. Revolving Spinal Prop. Its spinal lever accommodates the angle. A A, plates which revolve on screw posts, so as to fit the planes of the curve. These plates are curved to the form, and may be run up and down on the screw posts, to suit the height of the curve; they are a positive protection against breaking the prominent parts.

the hips, again ranging with the head and joints below, causing you to be perfectly erect, the shoulders still being, as in other experiments, the same as they were before erecting the body.

We think we have now shown to the satisfaction of the reader, that in order to straighten the form of a crooked person, or habitual drooper, *the support should never be placed higher than the hollow of the back.* And we earnestly entreat all who are interested in the matter of artificial support, either for themselves or others, to make the foregoing experiments, and follow their leadings.

We have said, that " a growing out " of the shoulder always attends a lateral curvature of the spine. This is because their relation to each other is simply that of cause and effect ; for, when the right shoulder is enlarged, and the left diminished, the spine between the shoulders is found bending to the right. This, at times, is so much the case, that it passes under the right shoulder blade, crowding it out of its place. The returning line of this curve, in its circular descent to the medial line of the back, forms a single spinal curvature. By following the course of this curve, it will be seen to cross the medial line of the back, describing as a natural consequence, a curve to the left, in the lower part of the spine, which, with the former curve near the right shoulder, forms what is called a double curvature. This lower curve crowds the left hip out of its place, in the same way that the upper curve resulted in an elevation of the right shoulder. By this means, also, the fulcrum is removed from the centre to the left side of the figure, whilst the serpentine course which the spine has assumed, materially diminishes the height of the body.

Spinal and Muscular Weakness.—We come now to speak of another modification of Spinal Affections, which may emphatically be termed Spinal and Muscular Weakness. This state has often been mistaken for inflammation, or genuine irritation of the spine ; a circumstance which is mainly attributable to the fact, that many of the sub-

BANNING TRUSS & BRACE CO., 704 Broadway, N. Y.

jects usually complain of symptoms peculiar to spinal irritation. The *error* is natural enough, but not the *treatment*. Spinal weakness is distinguishable from spinal irritation, by the absence of that intense tenderness which is usually felt in one or more of the spinal bones, and which causes the patient, on pressure, suddenly to cringe and shriek ; whereas, in mere spinal weakness, this pressure simply calls forth such moderate expressions as "it is tender," "it is sore," or "it aches." This distinction between spinal irritation and spinal and muscular weakness, is of the utmost importance in the treatment of spinal affections.

The proper treatment for the relief of these painful maladies consists in restoring the deranged or disordered mechanism of the system to its primitive relations, which, we have demonstrated, is the legitimate work of the Body-Brace, Spinal Prop, Centripetal Lever, or Abdominal and Spinal Shoulder-Brace. And, if we are not greatly mistaken, the cases, cited by us on page —, will satisfy the reader, that whilst diseases of the spine must forever resist all efforts at *compression*, they readily yield to the principle of *support*.

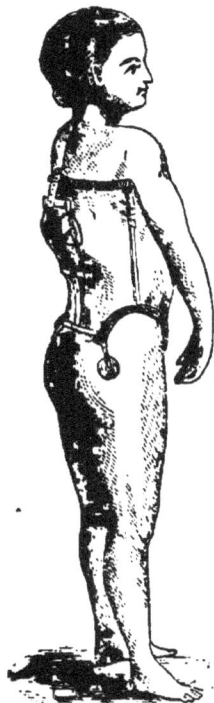

Fig. 19.

Fig. 19 represents the Revolving Spinal Prop. Immediately strengthening the whole person, and arresting caries and curvature. First, by upward support which converts the abdominal viscera into an internal brace. Second, by its crutch-like action, which holds the body's weight from the spinal curve. Third, by a strong drawing back of the shoulders by the caps on the shoulder-bow in front of the heads of the humeri; and, Fourth, by the strong bracing and pushing forward action of the revolving dorsal plates on the vertical screw rods upon the curvature. By revolving action, these plates are self-adjustable to any slope of the spinal angle at either side, with no necessity for any impingement upon the spinous protuberance. As the case improves, the vertical support may be successively increased by means of slides and screws in the side-posts.

BANNING TRUSS & BRACE CO., 704 Broadway, N. Y.

WHO SHOULD WEAR THE BRACE?

1st. All public speakers and singers who have any irritability of the throat, or an ungovernable and cracked voice, and whose vocal exertions are succeeded by languor and fatigue, with sense of *flatness*, faintness, or sinking at the stomach, and aching of the back.

2d. Who have dull pains and sense of oppression about the chest, with limited or hurried breathing on slight exercise, accompanied by short cough on attempting a *full* inspiration, and particularly where there is any predisposition to bleed at the lungs. Also, by *confirmed* pulmonics, as a very great temporary relief to the last. (See "Common Sense.")

3d. Who are troubled habitually with palpitation of the heart on slight exercise or excitement; and, also, a general nervousness, such as hysterics, lowness of spirits, gloom, causeless crying, religious hallucinations, hypochondria, melancholia, and temptation to suicide. All such as the above usually approximate the unnatural shape, and complain, more or less, of its general results.

4th. Who have any degree of dyspepsia, and complain of dull dragging pains in the sides, irritating affections of the stomach, liver and spleen, accompanied by a sense of deathly sinking or "*goneness*" at the stomach, aggravated on exercising, and on taking the *erect* posture; together with a sense of *shaking*, *heaviness*, and *pulling* or *dragging* on *walking* or *riding*.

5th. Who are habitually costive, have chronic diarrhea, or are in the latter stage of dysentery, and particularly all who are habitually afflicted with colic, and who have either *bleeding* or *blind piles*.

6th. Who are swag-bellied, have chronic peritonitis, and any variety of rupture of the bowels, or a predisposition to the same.

7th. Having affections of the prostate gland, or who have any difficulty in either retaining or evacuating urine, and particularly all of that unfortunate class of both sexes, who from various causes (whether they be personal folly or accident), are laboring under that seminal or genital weakness which, despite the best moral principles and all constitutional treatment, goes on to sap both body and mind of its strength, and too often terminates in insanity or imbecility. It cannot be too eagerly sought by young and old of both sexes, where the draining waste, the weakness of the loins and nerves, and the mental depression and weakness continue, after reforming from their follies. The effect is often wonderful on both body and mind.

8th. All who have any variety of spinal affection, whether it be a double or single curvature, or spinal irritation, or only *weakness*, attended by dull pains between the shoulders, or continual wrangling or grinding pains in the small of the back and hips, with coldness, numbness, swelling pains and varicose veins, with weakness of the lower extremities, particularly in chronic milk leg, and the hardening of flesh and ulcerations so consequent upon it.

9th. All pregnant ladies who are disposed to abortions, or suffer much in the last period of their *journey*, and particularly *every* lady immediately after delivery, and for weeks after (in such case should be applied over a wide linen bandage).

10th. All who have any of the above symptoms, in connection with any bearing down (falling of the womb), whites, painful, interrupted or profuse menstruation.

11th. All weak, lax-fibered, and rickety children, and youth who droop, lean and lounge, and who do not endure exercise well; and particularly young girls whose physical powers rather flag about the change to womanhood. If neglected *then*, they seldom fully recuperate afterward.

12th. Particularly those recovering from long confinement, by fevers, or any other cause, should use it—it will enable them to begin to move earlier and more among the convalescing influences; for sea-sickness, it is most grateful.

And lastly. All who are from habit or occupation disposed to droop, and are of a sedentary and rather delicate frame, should always *own one*, to use casually in walking, riding, or travelling, thus averting the consummation of their downward tendency.

All of the above maladies may be caused by the descent of

the internal organs, producing pressure from above on the one hand, and dragging on the other [re-read explanation], all of which the Brace may rationally relieve by supporting the back, and lifting the abdomen, thereby restoring a *natural* condition, and removing the physical power of discomfort. Reflect upon this before you say, " It cures too many things."

CAUSES OF THE DEBILITY OF LADIES, SO COMMON AT THE PRESENT DAY.—A neglect to support weak children. 2d. Neglecting to support fast-growing and lax-fibered girls, at the change from girlhood to womanhood. 3d. The present length of whalebone dress waists, and the enormous number and weight of skirts at this day commonly worn pressing downward.

HOW TO MEASURE FOR THE SELECTION OF ANY ONE OF THE BRACES.

1st. Number of inches around the body, two inches below the tips of the hip bones.

2d. Do. straight around the chest close by the armpits.

3d. Do. from each armpit to the tips of hip bones on each side.

4th. Height of the person.

HOW TO APPLY THE SEVERAL BRACES.

1st. Let every part on the shoulder, back and abdomen, be twisted to set flat and even.

2d. One thing always observe, viz.: Always have the front pad very low, and elevate the bowels with the hand, so as to have a *lifting*, and not a squeezing of the sunken organs.

3d. Commence to wear them loosely, and at short intervals at first, in all cases where they produce any pain or nervousness. A *tempered* perseverance will *always* overcome any physical or other inconvenience.

PARTICULAR ADVANTAGES OF THE BRACES OVER OTHER SUPPORT-ERS.—1st. They are cool. 2d. They are light. 3d. The pads can all be shifted up or down, right, or left as often as the spleen or necessity of the case may require. 4th. Their great and universal flexibility. 5th. They LIFT more than any other. 6th. Their back pads are four, and press on the weak hips, and particularly on the weak back, balancing and not restraining the body. 7th. The pads being of naked horn, stimulate and harden the muscles, while soft and cushioned ones relax and weaken, through heat and perspiration, besides giving them a rancid smell. 8th. They are so constituted as to admit of attaching to it any proper spinal apparatus, and also the most perfect pile and hernial trusses. 9th. They may combine with their mechanical influence the virtues of the galvanic battery, locally or generally applied

NOTES.

It is certain that a displaced or dragged condition of the vital organs may produce mechanical diseases, or greatly complicate and modify vital ones; since it is certain that the vital organs are as much under the law of a proper position and bearing as the bones are; hence why correct internal treatment too often stops short of a perfect cure, the Brace being requisite to mend the broken law of a natural bearing of the vital organs.

Is it not clear, that if the functions of fractured bones, or of dislocated joints, must be relieved by the nice readjustment of the same by *mechanical* means, then loss of function and irritation may also follow a dislocation of the vital organs; and that the latter must also be relieved by *mechanical* means, and not by medicine alone?

There are millions of votaries of fashion and pleasure, who, notwithstanding whalebones, corsets, and long bodices, are but poor languishing caricatures of good figures, and are unable to enjoy the very fashion they bow their necks to. These might secure the splendid form they desire, together with improved strength and grace, by the use of the *Brace*, which supports and braces the parts concerned; while the former only squeeze and compress them.

It is now established by experiment as well as reason, that the *Brace*, when judiciously applied, in cases of typhus fever, cholera, dysentery, diarrhœa, and other conditions where there is a tendency to fainting and collapse, immediately improves the pulse and strength, and removes the deathly sense of sinking and faintness (or goneness), by acting as a substitute for the strong abdominal muscles which should have kept the whole line of vital organs constantly packed or braced upwards in their proper places.

Thousands of invalids struggle with the proper symptoms of many acute and chronic diseases, but linger to *fully* recover, whom the Brace would immediately enable to rise and obtain the benefits of air, exercise, and society, as muscular debility, and a consequent loose and unsupported state of the vital organs, have alone retarded their complete recovery.

Said a distinguished clergyman of Vermont to the author: "Doctor, I feel persuaded that you are a louder and more extensive preacher of the gospel than any other man, for, judging from my renewed powers and faculties to preach, since wearing your Brace, and from what I have heard from other enfeebled clergymen who have experienced a similar relief, you will be instrumental in continuing many a successful messenger of peace in the pulpit, who otherwise must be silent."

The difference between Banning's Non-Friction Self-Adjusting Brace Truss for the relief and cure of Hernia is, that while the latter retains the Hernia by pressing upon the external ring and thus enlarging it; also giving undue pressure upon the spinal nerves and spermatic cord;—the former by its lifting

BANNING TRUSS & BRACE CO., 704 Broadway, N.Y.

action and non-friction movement calls for but slight pressure, not upon the ring, but the inguinal canal, causing a consequent contraction of the rings, and in many cases a *Speedy Radical Cure.*

Extract from a Patient's Letter.

"DESCRIPTION OF COMPLAINT.—Backache, terrible wrangling, twisting, pressing, bearing down or dragging in the groins; back and limbs very weak; dull, heavy pains in the lower part of my body; numbness and constant inward heat at the lower part of my abdomen; with prickling in my hips and down my limbs; inward heat and bearing down constantly very great; all of my disease lies in my back and lower abdomen. *Oh! how I long to see your Braces.*"

Thousands might use the same language.

Miscellaneous Cases.

The subjoined few cases are selected from many thousands of others, on account of their *brevity, comprehensiveness,* and expressiveness; several of an older date are among them, in lieu of others of a late date, purely because they occupy a smaller space, and are so tersely expressed.

From Sam'l Ross of New York.

DEAR FRIEND,—Some five years since, I made a brief statement of the great improvement of my son's case of spinal curvature, under thy treatment. I now have the pleasure to add that his health is perfect. His form is so much improved as not to attract particular notice; his strength and activity such as to enable him to run, jump, or wrestle, with impunity. So signal has been thy success with my son, as to compel me to advise all interested to resort to thee without delay.

SAMUEL ROSS, *Dentist*.

From A. Tanner of Warren, Pa.

For twenty-five years I was sorely afflicted with obstinate costiveness, piles, terrible sick headache, and melancholy, to such an extent that about one-third of my time I was unfit for corporeal or mental labor. I was soon *perfectly* relieved of them *all*, by the use of your Brace. Two years have now elapsed, and no return of my former troubles. I have also seen hundreds of similar cases equally relieved by the same means. I believe it would be *invaluable*, at least to all who are similarly afflicted.

ARCHIBALD TANNER.

From the Louisville Journal.

DEAR SIRS, I feel it to be due you as well as the community, to say, that although I have worn your Brace for relief of piles, only a short time, I have already experienced more complete and speedy relief from the most painful suffering of twenty years than I had ever anticipated or hoped; and with great pleasure, I would recommend your agreeable and effective Brace to those similarly affected.

Mrs. J. N. McMICHAEL.

From the Louisville Journal.

Mr. PRESTON:

I wish through you to say to the suffering public, that after but a few days trial of the Banning Pile Brace, I have experienced inexpressible relief from most painful and exhaustive bleeding piles and general debility. My improvement in posture, inward strength, and ability to preach, is very great. The relief is perfect.

W. W. SELLERS.

Extract from a Letter of a Lady to her Brother,
a Physician.

DEAR BROTHER,—* * * * * After trying various other supporters to no avail, I was induced by a friend to try one of the Banning Body Braces; have worn it some time, and can testify that language is inadequate to express the relief I have experienced from it. That constant dull pain in the right side is wholly removed by its use; likewise the pressing, bearing down of the abdomen; the piles and constipation of the bowels are all relieved. You well know how difficult it has been for me for years to walk any distance; but since I have worn the Brace I have walked miles daily with slight inconvenience. I could not believe anything could have been devised that would so perfectly support the body. And now, by dear brother, I want you to recommend this most useful instrument to all your patients afflicted as I have been.

M. K. EVERETS, *New York.*

From N. McConaughty, Millville, N. J.

DEAR SIR,—I have worn your Brace fourteen months, and am a rescued man. Seven years of constipation, piles, nervous derangement, and general debility, are forgotten as a dream. I own my life to your Brace. No money would induce me to part with it. God surely sent you to me in my utter extremity. May He send you everywhere, the apostle of humanity and health.

Yours gratefully,
N. McCONAUGHTY,
Pastor Presbyterian Church, Millville, N. J

From J. W. Wiley, Pennington, N. J.

I have worn your Brace for several months, on account of abdominal weakness, admitting of a drooping of

BANNING TRUSS & BRACE CO., 704 Broadway, N. Y.

it indispensable to preachers and public speakers who are laboring under abdominal weakness, or a loss of expulsive force in speaking. To all such I cordially recommend this instrument.

J. W. WILEY.

From Lucretia Lambert, Allowaystown, N. J.

DEAR SIRS, After wearing your Brace for one year, the most extreme and painful female (or uterine) weaknesses have vanished. For ten years I was unable to walk or ride without intense suffering. Thousands had been spent upon the best physicians and supporters, without avail. But for your Brace I would not accept of this world full of money in exchange.

I pray you devote your whole life to the treatment of similar sufferers.

LUCRETIA LAMBERT.

From Geo. Hitchins, Millville.

DEAR SIRS, I am happy to say that the Brace you applied to my daughter has not only restored her strength, but also removed every vestige of her double lateral curvature of the spine and ungraceful inequalities of her hips and shoulders.

I think all parents of weak and fast growing or deformed children should consult you at once.

Your obedient servant,
GEORGE HITCHINS,
Pastor First M. E. Church, in Millville.

From that old Veteran, Commodore Charles Ap C. Jones.

NEAR PROSPECT HILL, VA., *Oct.* 8, 1846.

DEAR SIRS, Ever since 1814, when I received a gun-shot wound (the ball still remaining in my body), I have suffered intensely from pain in my back; so, at times, as to prevent riding or walking without destroying all comfort. I have tried the various belts and supports, but found the remedy worse than the disease.

Since wearing your Spinal and Abdominal Brace, I have gone through some severe and protracted exercise without rest. On one occasion, I was for eleven hours either on my feet or in the saddle, without the least inconvenience then or afterwards. But for the Brace I could not have accomplished a tithe of this without intense suffering then and for days afterwards.

I am certain that my general health also has been greatly renovated by your Braces, and I wish you great success in your pilgrimage for the relief of suffering humanity.

Gratefully yours,
Thos. Ap C. Jones,
U. S. Navy.

Extract of a Letter from the Rev. David Caldwell, Rector of St. Paul's Church, Norfolk, Va.

I am now truly rejoiced to hear you are coming South. I hope you will find the trip one of profit to yourself, for I am sure it will be one of benevolence to others. If you wish, I would write you out an accurate description of *how great* service your Brace has been to me, especially as a *stay and support* in speaking. If I could not procure another, I would not part with it for *ten times its weight in gold.* I hope you will, by all means, visit Norfolk. I know many who need your Brace, and if they only knew its excellence, would furnish themselves with it. I am determined to keep a second Brace always on hand, provided against any accident to the first. I have been repeatedly urged to write to you for several of your Braces—and although wholly adverse to engage in secular matters, yet I regard your "Body Brace" *such a blessing to individuals*, that if you will send me one dozen by Mr. ———, or by the weekly packets between this port and New York, I will be responsible for them.

David Caldwell.

From Chas. S. Pope of Washington, D. C.

Before I commenced to wear your Abdominal and Spinal Shoulder Braces, I was helpless from bleeding piles, bleeding at the lungs, indigestion, constipation, and great soreness, "goneness" at the stomach. Between bodily pain and mental distress, my hope was "forlorn."

But, from the same day on which I applied your Braces, both mental and physical distress were diminished, and in a few days every one of my sufferings had vanished, and I have now increased many pounds in weight, and am able, irrespective of weather, to attend to the most laborious duties. It is with a transport of desire that I wish multitudes to apply your agreeable Braces for similar weaknesses.

CHARLES S. POPE.

CURE OF TOTAL PARALYSIS.

From Wm. Springsted of Rochester, N.Y.

SIRS, When my wife came under your treatment and applied your Spinal Prop, she, besides the severest uterine weaknesses, had for six months endured *total paralysis* of both limbs, without power to make the slightest movement, even of her toes; but immediately after, her uterine symptoms were relieved, and within four weeks she could move her limbs and stand; and now, in three months, she does her own housework. So wonderful is the result, and so great my desire to have all spinal and uterine cases apply to you, that it seems as if the stones would cry out were I to hold my peace.

Ever your ob't serv't,
WM. SPRINGSTED.

From Mrs. Abbie Potter of Warren, Trumbull Co., O.

For ten years I have been confined to my bed with spinal curvature, falling of the womb, incontinence of uterine, terrible constipation, and extreme indigestion and emaciation, and loss of taste in my mouth. For seven years I was unable to hold a pen, and at no hour of this time was I kept from attracting the attention of passers in the street by my constant moans, without

heavy doses of morphine. But, I am redeemed, and restored to both strength and health, by your simple abdominal
and spinal brace. People come many miles to be "certain"
that the miracle really has been performed. Could weak ladies
know half of what your Brace has done for me, each one would
hasten to possess it.

Ever gratefully yours, etc.
ABBIE POTTER.

*From Mr. Robt. Montgomery of Youngstown,
Mahoning Co., O.*

My wife, who was for four years the subject of paralysis, of
one leg and arm, and of great uterine trouble, never being able
to sit up or to allow her feet to hang down, is most strangely
restored to health and the free use of her limbs, by the simple
application of your Brace, and this, after all doctors and remedies had left us forlorn. And when I see that the result has
been as philosophical as it is grand, I yearn to get your work
on Mechanical Support into the hands of every family, and your
Brace upon their weak or deformed bodies.

Ever your obedient servant, etc.
ROBT. MONTGOMERY.

From Mrs. H. Miller of Pittsburg.

Under God, I owe you or your Brace everything, it
having utterly cured me of the most terrible instance of milk
leg, with varicose veins and enormous swelling and ulceration I
ever heard of. It had been of seventeen years' duration, and
amputation had been advised; shortly after the Brace lifted
the weight of my bowels from the veins of my limbs, the swelling, hardness, and blackness, left, and in six weeks the ulceration was gone. I, now, just for the pleasure of the thing, visit
the market every morning.

God bless you,
Mrs. HARRIET MILLER.
H. M.

BANNING TRUSS & BRACE CO., 704 Broadway, N. Y.

BROOKLYN, *October* 24, 1865.

From happy experience in my family, I cannot doubt that your Uterine Balance will, in your own hands, cure the most protracted and extreme Retroversion of the Uterus.

C. McKEAN, 158 *Fulton avenue.*

———

WISCASSETT, ME., *October* 24, 1865.

SIRS, Your Uterine Balance having cured in my family an extreme case of Uterine Anteversion of eleven years' duration, which resisted the best skill in Europe and America, I desire to inform all feeble ladies of the fact, and encourage them to make application to you.

SAMUEL D. DOANE.

———

BROOKLYN, N. Y., *October* 20, 1865.

DEAR SIRS, Your Uterine Balance has given unsurpassed relief to a case of Uterine Retroversion in my family, after thir-teen years of failure under the most distinguished Physicians.

CHAS. H. ZUGALLE,
Clinton ave., near Gates.

———

From Medical Profession of Pittsburg.

We, the undersigned, having used in our practice " Banning's Patent Brace," for the relief of cases of similar prolapsus uteri, cheerfully testify to its being the best instrument we have met with to fulfil all the indications required in the case, which can be required from an external support.

A. N. McDOWELL, M. D.
T. F. DALE, M. D.
JOS. P. GAZZAM, M. D.

From Rev. Jos. McElroy, D. D., Pastor Scotch Presbyterian Church, New York.

MY DEAR SIRS, I have been accustomed to public speaking as a Minister of the Gospel for half a century, and in the earlier part of my ministry I suffered considerably in the lower part of the abdomen after every time I preached. I tried various remedies, thinking that my suffering proceeded from weakness in that portion of my system. I used for a time the Russian Belt (as it is termed), but with no good result. Some twenty-two years ago, happening to see a report of one of your lectures in the "Tribune," I called on you, procured one of your "Braces," and from that day to this I have not been an hour without it, except when in bed. It has been of unsurpassable advantage to me—advantage, not only in *speaking*, but in *walking*. In both it imparts a buoyancy to the system. It produces, however, one peculiar effect which I wish to mention to you: I have found by experience again and again, that when I have put it on slack and loosely, not only has a disposition to void urine more frequently than usual followed, but diarrhœa has set in, and by simply tightening the brace, both these effects have been checked. The philosophy of this I know not, but I do know the fact.

You are perfectly at liberty, my dear sir, to make what use you please of this communication.

Truly,

J. McELROY.

From Ex-Governor Wright of Indiana, formerly American Minister to Berlin.

I most fully agree with the statements of Doctor McElroy, from a personal trial in public speaking for years, and believe all the Doctor has stated, as I have suffered for years with Diarrhea.

J. A WRIGHT.

From Professor Mott, of New York, and others.

The undersigned have examined your novel views on the mechanical pathology of many affections of the viscera, and believe them to be highly interesting and worthy the serious attention of the Medical profession.

> VALENTINE MOTT, M. D.,
> ALEX. B. WHITING, M. D.,
> J. KEARNEY RODGERS, M. D.

ST. GEORGE'S RECTORY, *Jan.* 26, 1854.

DEAR SIRS, I have read with much interest your whole book, but especially your Lecture on the human voice. I consider your principles, in reference to the latter, entirely sound and practical, and should rejoice to have all our public speakers put them in thorough practice, both for their own health, and for the efficacy of their labors upon others.

> Your friend and servant,
> STEPHEN H. TYNG.

Your views on spinal curvatures and their derivative affections, appear incontrovertible, and his various instruments and appliances for the cure of these maladies are unsurpassed for their elegance, lightness, mechanical ingenuity, and scientific adaptation to the ends contemplated in their employment.

> J. T. CURTIS, M. D.,
> 28 *West Fifteenth street.*

NEW YORK, *June* 10, 1854.

NEW ORLEANS, LA., *July*, 1870.

DEAR SIRS, Though I have worn your Truss only thirteen months, I am able to say that I am *radically cured* of a Double Rupture of immense size.

Truly yours,

GEO. McGIFFIN.
Cor. Camp and Canal sts.,
New Orleans, La.

———

COLUMBUS, GA., *Jan.* 15, 1870.

It gives me pleasure to testify as to the remarkable efficacy of your Truss in the case of my little boy Eddie. Some twenty days ago, when, by the advice of my physician, I brought him to you he was afflicted with a Hernia of considerable size, and with which for the last two years he had suffered considerably. From the day your Truss was applied he has been able to exercise with comfort, and yesterday, when the Truss was removed, the Rupture did not descend. Please write me as to whether he shall continue wearing the Truss, and if so, for how long a period.

Yours gratefully,

A. G. HENDRICKS.

———

SAVANNAH, GA., *Nov.* 30, 1869.

It gives me pleasure to be able to bear witness as to your skill as a physician, as also to the intrinsic merit of your Umbilical Rupture Truss. My boy, aged seven, had an Umbilical Rupture of enormous size. It affected his health so seriously that we had no hopes of his living the year out. You not only dexterously reduced the Rupture, but under your treatment he has so far recovered as to give us no fear as to his

BANNING TRUSS & BRACE CO., 704 Broadway, N. Y.

life, and the Rupture itself, to use his own language, "hardly comes out." I examined it the other day and found that I could not put the point of my finger in the rent. Before he came under your treatment, the sack and contents would have filled a peck measure and the entire fist could have been thrust into the rent.

Yours truly,

C. C. C. BRUEN.

———————◆———————

To the Profession.

Appliances for all Deformities, Debilities or Deficiencies of the Human Body.

The most approved apparatus for Spinal Curvature, Club-Foot, Hip Joint Disease, Bow Legs, Weak Ankles, Knock Knees, Paralysis, Wry Neck, Pott's Disease of the Spine, Muscular Contractions, Fracture of the Patella, Splints for Ununited Fractures and Resections. Especial attention given to devising and constructing Instruments for peculiar or unusual cases.

BANNING TRUSS & BRACE CO.,
No. 704 Broadway, N. Y.

To the Profession.

Appliances for all Deformities, Debilities or Deficiencies of the Human Body.

The most approved apparatus for Spinal Curvature, Club-Foot, Hip Joint Disease, Bow Legs, Weak Ankles, Knock Knees, Paralysis, Wry Neck, Pott's Disease of the Spine, Muscular Contractions, Fracture of the Patella, Splints for Ununited Fractures and Resections. Especial attention given to devising and constructing Instruments for peculiar or unusual cases.

BANNING TRUSS & BRACE CO.,

No. 704 Broadway, N. Y.

(OVER)

LIST OF ILLUSTRATIONS.

THE BANNING

Truss and Brace Company's

IMPROVED SURGICO-MECHANICAL APPLIANCES

So imitate nature as to lift, brace, expand and rest depressed, contracted and overtaxed parts, and do away with the necessity for their own permanent use. While allowing perfect freedom of motion, they do not show upon the wearer, and immediately improve the figure of deformed drooping, or disfigured children, young ladies and others. They possess a combination of valuable properties to which no other system of supports pretends to lay claim, and which the experience of over thirty years enables us to offer to the profession and the public with increasing confidence.

☞ ALL communications, whether on professional or other business, must be addressed to the Banning Truss and Brace Co., at 704 Broadway, (NO OTHER OFFICE OR ADDRESS,) and all Drafts and Remittances must INVARIABLY be drawn to the order of the TREASURER of the Company.

www.ingramcontent.com/pod-product-compliance
Lightning Source LLC
Chambersburg PA
CBHW022001190326
41519CB00010B/1350